spot

EXTREME WEATHER

HURRICANES

by Anastasia Suen

AMICUS | AMICUS INK

storm surge

boards

Look for these words and pictures as you read.

eye

trees

Look at that spinning storm.

It is a hurricane!

Hurricanes start over the ocean.
The water is warm.
Warm air rises.
Clouds form and spin.

See the boards?
They cover the windows.
The glass won't break.

boards

eye

See the eye?
It is the middle of the storm.
It has no wind or rain.

trees

See the trees?
The wind is very strong.
Trees and buildings can break.

storm surge

See the storm surge?
Wind pushes the water onto land.
It floods the town.

The storm goes away.
People clean up.

storm surge

boards

Did you find?

eye

trees

Spot is published by Amicus and Amicus Ink
P.O. Box 1329, Mankato, MN 56002
www.amicuspublishing.us

Library of Congress Cataloging-in-Publication Data
Names: Suen, Anastasia, author.
Title: Hurricanes / Anastasia Suen.
Description: Mankato, MN : Amicus, [2021] | Series: Spot. Extreme weather | Audience: Ages 4-7. | Audience: Grades K-1.
Identifiers: LCCN 2019036318 (print) | LCCN 2019036319 (ebook) | ISBN 9781681519449 (library binding) | ISBN 9781681525914 (paperback) | ISBN 9781645490296 (pdf)
Subjects: LCSH: Hurricanes—Juvenile literature.
Classification: LCC QC944.2 .S84 2021 (print) | LCC QC944.2 (ebook) | DDC 551.55/2—dc23
LC record available at https://lccn.loc.gov/2019036318
LC ebook record available at https://lccn.loc.gov/2019036319

Printed in the United States of America

HC 10 9 8 7 6 5 4 3 2 1
PB 10 9 8 7 6 5 4 3 2 1

Alissa Thielges, editor
Deb Miner, series designer
Ciara Beitlich, book designer
Aubrey Harper, photo researcher

Photos by Shutterstock/MISHELLA cover, 16; iStock/johnnorth 1; Alamy/Aflo Co., Ltd. 3; iStock/Maryna Patzen 4–5; Getty/CHUYN 6–7; NASA/Alexander Gerst 8–9; Getty/Bobbushphoto 10–11; iStock/Fouque 12–13; Alamy/Richard Green 14–15